SOLUTIONS MANUAL

AUTOMATED PROCESS CONTROL SYSTEMS

SECOND EDITION

CONCEPTS AND HARDWARE

RONALD P. HUNTER

A Division of Simon & Schuster
Englewood Cliffs, N.J. 07632

ISBN: 0-13-054487-6

Printed in the United States of America

Chapter 1

1. It stresses the systems applications and effects of the components and circuits rather than those components and circuits themselves.

2. Refer to portions of "Purpose of Automatic Control."

3. Because of increased reliance on automated industrial production systems, automated communications systems, automated commercial systems, and automated consumer systems. Automation affects nearly every facit of modern life and its influence and applications are increasing at a fantastic rate.

4. Extension of answer to #3 above.

5. Refer to appropriate portions of "How an Industrial Control System is Implemented."

6. Individual example. Make sure to identify feedback loops.

7. Refer to section titled "Goal of Automatic Control Theory."

8. Due to the system effects of negative feedback which result in the ability to exercise automatic quality control over the end product as system parameters vary, whereas no automatically correcting control is possible with open-loop control.

9. Refer to appropriate portions of "Introduction to Automatic Control Theory"; make sure all key concepts are included.

10. Optimum control of time-variant processes. Refer also to the section on "Goal of Automatic Control Theory."

11. To provide the controller with knowledge concerning the current state of the process so that it can make its preprogrammed calculations to decide if corrective action should be initiated (or sustained) and if so, how much, in what direction, and how fast (at what rate).

12. Refer to the definition of Feedback in the glossary.

13. Referring to the error signal (difference between setpoint and the measured feedback variable); with negative feedback the corrective action (if any) will cause the error signal to be decreased in magnitude, whereas positive feedback will cause the error signal to increase in magnitude until the system saturates.

14. The error signal (also called the actuating error signal or deviation signal) is the simple difference between the setpoint and the measured process control system feedback variable (both being properly scaled for the comparison). It is the input upon which the controller bases its control decisions.

15. The desired condition of the process variable being controlled.

16. It is the portion of the control system which receives the setpoint and the measured process variable (feedback) and determines the magnitude and polarity of the difference between the two.

17. See the definition of Controller in the glossary.

18. To provide for all the system's signal conditioning requirements (phase inversion, amplification, magnitude conversion, voltage-to-current and current-to-voltage conversion, filtering, impedance matching, etc.).

19.

Closed-Loop	Open-Loop
Feedback	No feedback
Self correcting systems	No self correction possible
Can be extremely complex	Normally fairly straight forward
Can become unstable	Inherently stable
Used on both analog and digital systems equally	Used primarily on digital (on/off) systems

20. See definition of Controller in glossary. It is the hardware necessary to implement the control strategy; process the actuating error signal (deviation) based upon the control strategy, and develop the necessary output command signal.

21.

Closed-Loop	Open-Loop
Automatic setpoint control possible	No setpoint control possible
Used with time-variant and batch processes	Used with batch processes primarily

See also the answers to question #19, most of the comments there also apply to this question.

22. Refer to figure 1-6.

23. Other than empirically (which works on smaller and more familiar systems), the only technique available to process control engineers to accomplish process simulation and initial controller design is through the mathematical modelling of the process; and this leads directly to a mathematical model of the controller. The mathematics necessary to accomplish and understand this process and its results involve calculus and statistical theory.

24. Processes whose output is manufactured in discrete incriments or batches, as opposed to being manufactured on a continuous basis.

25. The term batch process can be specifically defined as compared to continuous processes. A continuous process is one whose time-variant parameters are measured and controlled based primarily on their time-varying characteristics. Batch processes are processes in which the end product is manufactured in discrete batches, ie. a batch of cement, a batch of chemicals, a batch of pastery dough. The system variables in batch processes are controlled more based upon the quantity of the raw material to be mixed than on the time rate of mixing. The term is complicated further when it is understood that certain portions of most batch processes include continuous portions, ie. cooking a premixed batch of material according to some predefined time-temperature relationship.

26. The problem of maintaining adequate system stability under all sets of operating conditions.

27. Feedback 28. B. 29. E. 30. D.

31. E. 32. True 33. E. 34. B.

Chapter 2

1. Refer to the definition of Transducer in the glossary.

2. Refer to the definition of Sensor in the glossary.

3. A sensor is a transducer which is connected into a system such that it provides an input to a controller. All transducers are not necessarily sensors, all sensors ARE transducers.

4. The sensing of gas tank level, oil pressure, speed and engine temperature to aid the driver in properly controlling an automobile.

5. By types of input variables measured, by types of output provided, and by whether they are analog or digital in nature.

6. See page 4 in this manual.

7. Refer to the definitions of the terms Analog and Digital in the glossary. Those definitions include the comparison. A further comparison is made in the introductory paragraphs in Appendix B.

8. Refer to the definitions of Absolute and Incrimental in the glossary. Those definitions include the comparison. Refer also to appropriate portions of "Linear Motion Sensors" for this comparison.

Problem #6, Chapter 2

Sensor	Variable Controlled	Principle of Operation	Advantages	Disadvantages
Potentio-meter	Position	position-to-var R	simple, can be made non-linear, good output, cheap available, effective	mechanical & environ-mental problems
Inductive	Position	position-to-var inductance	rugged, sensitive, elect-rical isolation	changes in inductance difficult to measure expensive
Capacitive	Position	position-to-var capacitance	rugged, sensitive	changes in capacitance inconvenient to measure
Digital Encoders	Position	position-to-digital	sensitivity, fixed accuracy, no drift, digital output, large range, effective	expensive, needs other electronics
Synchro	Position, Motion	position-to-AC	rugged, electrical iso-lation, effective	AC output inconvenient, mechanical problems
Various	Rotary Motion	optical, magnet-ic, mechanical	some solid state & non-contact, electrical iso-lation, inexpensive, digital output	digital outputs only
Various	Proximity, Limit	optical, magnet-ic, mechanical	simple, inexpensive, rugged, effective	digital outputs only
Strain Gage	Position (Force & Torque)	strain-to-var resistance	inexpensive, rugged, effective	electrical output inconvenient to measure
Piezo-electric	Changing Force	distortion-to-charge	sensitive, inexpensive, rugged	only force changes, charge amp rq'd

9. From the incrimental position encoder, two wires, where each digital pulse will signify a motion equivalent to one incriment of resolution of the sensor's scales. Another pair of wires will normally, but not necessarily, be included with a digital signal on them indicating the direction of the incrimental change in position. From an absolute position encoder there will be as many (pairs) of wires as there are bits of resolution. When these wires are simultaneously looked at, the binary output word will directly relate to a unique position on the sensor's fixed scale (within the resolution of the sensor).

10. Motion, as used in this chapter, is defined as change in position.

11.

<u>Analog</u>	<u>Digital</u>
(advantages)	(disadvantages)
Compatible directly with process	Requires A/D & D/A conversion
"Infinite" resolution	Discrete limit to resolution
Large selection available	Limited selection available
(disadvantages)	(advantages)
Due to drift requires recalibration & zeroing	Normally no drift problems
Must be converted for use with digital systems	Directly compatible with digital computers
Displays require interpretation	Displays formatted for direct readout
Limited availability of storage of info	Virtually unlimited permenant storage capabilities
---	Many types of displays available not possible with analog

12. Refer to definitions of the terms Resolution, Sensitivity and Accuracy in the glossary. The definitions of these terms are implied in this chapter.

13. Refer to definitions of the terms Range and Span in the glossary.

14.

<u>Electrical</u>	<u>Mechanical</u>
(advantages)	(disadvantages)
low maintenance	significantly higher maintenance
long life	limited life due to wear
high frequency response	limited frequency response
measurement takes little power from process	normally takes a significant amount of power from process

The only advantages of mechanical transducers over electrical transducers may be in their availability, and in some cases their power output when used as actuators.

15. Nearness or closeness to.

16. Position sensors are basic sensors, force and torque are measured indirectly by position sensors; where the position sensors actually sense the distortion in a calibrated shaft or column. The amount of distortion is proportional to the torque or force being measured.

17. A.	18. A.	19. True	20. C.
21. B.	22. D.	23. C.	24. False
25. E.	26. E.	27. A.	28. E.
29. B.			

30. A change in voltage (or charge) for a change in pressure (or distortion).

31. C.	32. True	33. B.	34. C.
35. D.			

36. An incrimental linear or rotary position encoder.

Chapter 3

1. Refer to the definitions of Hydraulics and Pneumatics in the glossary.

2. Pressure, level and flow.

3. Refer to figure 3-2 and the text material supporting that figure.

4. Absolute = Gage + Atmospheric. Absolute is a measurement with respect to the vacuum of outer space. Gage is a measurement with respect to ambient. Gage reads zero when open to atmospheric (ambient) pressure, and compares other pressures to that reference.

5. Differential pressure isn´t absolute, gage or atmospheric; it is a measure of one variable pressure as compared to another variable pressure. It only responds to the difference between them.

6. Refer to figures 3-4, 3-5 and 3-7; and appropriate portions of the section on "Pressure Sensors".

7. Refer to figures 3-8 and 3-9; and appropriate portions of the section on "Pressure Sensors".

8. Since most pressure sensors transduce pressure to mechanical motion (or distortion), any of the mechanical-to-electrical transducers can be used to transduce the pressure measurement to electrical.

9. Refer to figures 3-14 through 3-21; and appropriate portions of the section on "Level Sensors".

10. Since most fluid flow sensors actually respond to flow velocity, the most basic problem is sensing the average velocity across the cross sectional area of the pipe. The flow velocity profile across the pipe is nonlinear and is affected by other variables in addition to the quantity of flow.

12. Refer to figure 3-25 an appropriate portions of the section on "Flow Sensors".

13. Refer to figure 3-26 and appropriate portions of the section on "Flow Sensors".

14. Refer to page 8 of this manual for this answer.

15. A.	16. Absolute	17. "Zero"	18. E.
19. C.	20. B.	21. A.	22. D.
23. E.	24. D.	25. C.	26. D.
27. E.	28. A.	29. C.	30. B.
31. D.	32. D.	33. E.	34. D.
35. C.	36. A.	37. B.	38. D.
39. True	40. False	41. D.	42. E.
43. C.	44. B.		

Chapter 4

1. Refer to definitions of the terms Heat and Temperature in the glossary; and to appropriate portions of the introduction to this chapter.

2. There are no practical (true) direct heat measurement sensors.

3. By measuring temperature and mass, and subsequently calculating heat content using those measurements.

4. Change in physical volume of liquids and gasses, and physical dimensions of solids with temperature change; change in electrical resistance of wire and semiconductor materials with temperature change; the Seebeck effect (thermocouples); the fact that heat is a radiant form of energy and can be measured remotely by radiation and optical techniques.

5. Refer to the definitions of the terms Thermistor and Thermocouple in the glossary.

6. Refer to the definition of the term Thermostat in the glossary, to figure 4-4, and to appropriate portions of the section on "Bimetalic Temperature Sensor".

7. The fact that heat is a radiant form of energy and can be sensed remotely by radiation and optical techniques.

8. Refer to figures 4-10,4-13, and 4-14, and the section on "Pyrometers".

Problem #14, Chapter 3

Sensor	Variable Controlled	Principle of Operation	Advantage	Disadvantage
Manometer	Pressure	pressure/height of liquid	simple, no calibration, rugged, sensitive, effective, diff. press.	not good for large press differences, difficult to interface in systems
Bourdon	Pressure	pressure deforms shaped tube	rugged, reliable, large or small press, cheap	mechanical problems, limited sensitivity
Bellows	Pressure	pressure deforms corrugated tube	sensitive to small pressures, rugged	expensive, limited range. small output
Diaphram & Capsule	Pressure	pressure deforms stamped discs	sensitive to small press, rugged, reliable, cheap	limited ranges, small output
Various	Level & Limit	float, pressure, nuclear, ultra-sonic, heat, optical, conduct.	good selection of effective & reliable devices	some are quite expensive and limited in applications
Nutating Disk	Flow Quantity	volumetric flow	direct volumetric flow measurement	mechanical, limited range, expensive, high impedance, limited applications
Orifice, Nozzle, Venturi, Pitot	Flow Velocity	velocity causes pressure drop across restriction	relatively inexpensive, effective	high impedance, limited ranges, limited applications, requires pressure sensor
Variable Area	Flow Velocity	pressure drop across float	visible display, accurate, relatively inexpensive	difficult to interface to, limited range&sensitivity, high impedance, clean fluids
Turbine	Flow Velocity	turbine wheel	accuracy, works with petroleum products	clean lubricants only, high impedance, maintenance, expensive
Magnetic	Flow Velocity	conductor moves thru magnetic field	no obstructions to flow, wide rangeability, good sensitivity, no wear, accurate, chunks OK	very expensive, exotic electronics, conductive fluid
Fluidic	Flow Velocity	wall attachment oscillator	simple, wide rangeability low impedance	clean fluids
Ultrasonic	Flow Velocity	vortex shedding & doppler	no special piping, any fluid, accurate, digital	new techniques

Problem #9, Chapter 4

Sensor	Variable Measured	Principle of Operation	Advantage	Disadvantage
Filled systems	Temperature	thermal expansion to pressure	simple, reliable, rugged, inexpensive	sensitivity, response speed, sealed system, temperature ranges
Bimetalic	Temperature	linear expansion of metals	simple, reliable, rugged, inexpensive	sensitivity, response speed, temp. ranges, mechanical output
RTD & RTB	Temperature	variable resist-ance	simple, reliable, rugged, sensitive, replacability	small changes in electrical output
Thermistor	Temperature	semiconductor variable R	simple, reliable, rugged, extremely sensitive, small	difficult to maintain manufacturing tolerances
Thermocouple	Temperature	e.m.f. generator	simple, reliable, rugged, sensitive, fail safe, wide rangeability	small electrical output, compensation & reference required
Pyrometers	Temperature	radiation (infrared or optical)	measures very high temperatures	only for very high temps, expensive, not rugged, manually operated

Problem #10, Chapter 5

Sensor	Variable Measured	Principle of Operation	Advantage	Disadvantage
Photocond-uctive	Light	var. resistance	small, sensitive, inexpensive	none ?
Photo-voltaic	Light	emf generator	small, sensitive, inexpensive	low power output
Photo-emissive	Light	emission of electrons	sensitivity	expensive, limited to high voltage systems
Radiation & Thickness	Radiation, Thickness	ionization & scintillation	sensitivity, analog or digital outputs	high voltage, expensive, not rugged
Humidity & Moisture	Humidity, Moisture	evaporation, hair & chemical sensitivity	sensitivity, accuracy	expensive
Timers	Time	R-C network, synch, motors	rangeability, accuracy, inexpensive, reliable	some are mechanical, component drift
Counters	Various	contact, optical, magnetic	inexpensive, reliable, accurate	none ?

9. See page 9 (upper half) in this manual.

10. When two dissimilar metals are joined together an electrical potential (emf) will be generated across the junction. This emf will be proportional (within certain fixed ranges) to the temperature of that junction.

11. B. 12. False 13. D. 14. E.

15. A semiconductor device doped to increase its temperature sensitivity.

16. Two wires, made of different alloys, are simply welded together on one end.

17. B. 18. E. (Refer to Appendix A also) 19. C.

Chapter 5

1. Refer to the definition of Light in the glossary.

2. Photoemission, photoconduction and photovoltaic. Refer to appropriate portions of the section on "Light Sensors" for descriptions.

3. X-radiation and nuclear gamma radiation.

4. Refer to figure 5-7 and the text material which supports it.

5. It is a percentage expressing the amount of moisture actually present in the air as compared to the maximum amount of moisture that air under identical temperature and pressure could hold before precipitating out.

6. Due to the cooling effect of evaporation from the wet bulb.

7. Time.

8. Because most automatic process control systems are controlled based upon the time variation of their parameters; therefore a time reference is absolutely necessary.

9. A series of wheels geared together with 10:1 reduction ratios between each.

10. See page 9 in this manual (the bottom half).

11. Refer to appropriate portions of the section on "General Comments on Selection of a Sensor".

12. D. 13. Photoemissive 14. Geiger-Muller tube
15. C. 16. E. 17. B.

18. Linear measure, rotational speed, objects moving along an assembly line, the quantity of events occurring during some interval (batch processing), etc.

19. C.	20. E.	21. B.	22. Gamma ray
23. D.	24. D.	25. E.	

Chapter 6

1. Provide the negative feedback signal; limit control system time response characteristics, overall sensitivity and accuracy; provide the means for monitoring the performance of the process.

2. A statement of how close a measurement is to the "TRUE" value.

3. Describe figure 6-4 in words.

4. Standards are measurement quantities whose values are accepted as being the "TRUE" value; against which other measuring instruments are calibrated.

5. To determine (and adjust) the accuracy of an instrument.

6. Refer to the introduction of this chapter for reference.

7. A. Does not account for wind, temperature, pressure.
   B. Same RPM can be required at different speeds in different gears.
   C. Varies with other parameters also.

	Range	Span
8.	0 - 100 MPH	100 MPH
9.	0 - 99,999 Miles	100,000 Miles
10.	100 MPH - 700 MPH	600 MPH

11. (85 - 5) / 8 = 10. MPH/Inch.

12. Span = (110 + 30) = 140.
    Distance = (6) (pi) (3/4 assuming we are only using 3/4 of the circular scale) = 14.14 inches.
    Scale factor = 9.9 F^{o}/Inch.

13. Range = 0-100% in x, 0 -100% in y.
    Span = 100 in both x and y.
    Linearity = maximum deviation between actual and ideal.
    Scale factor = 100% / (length of the scale).

14. A. A VOM is a shop and test instrument.
    B. A standard cell can be either a primary or a secondary standard.
    C. A = primary standard.
       B = Reference or secondary standard.
       C = Reference or secondary standard.
       D = Reference or secondary standard.

15. C.   16. D.

Chapter 7

1. Impedance matching connotes the careful consideration of the interconnection of various devices, making sure each successive device connected in a system will not unduly (adversely) affect the overall accuracy, sensitivity, time response, etc. of the preceding device by loading it down or introducing noise. Improper impedance matching unnecessarily degrades overall system performance by degrading parameter measurements.

2. Voltage
   Current - normally converted to voltage and processed
   Resistance
   Capacitance - normally connected as frequency determining elements in oscillating circuits whose outputs are processed as frequency measurements
   Inductance - same comment as for capacitance
   Frequency

3. Sensitivity to battery voltage
   Manual recalibration required prior to each measurement
   High measurement currents through unknown resistance
   Sensitivity to lead resistance (and variations in lead R)
   Poor sensitivity to small variations in relatively larger values of resistance

4. Rephrase the section on "Wheatstone Bridge Measurements", referring to figure 6-4.

5. The balancing potentiometer is a precision piece of equipment, and therefore very expensive. Also it will exhibit wear, and thereby loose its precision and accuracy as a function of use.

6. Rephrase the section on "Current-Operated Bridge Measurements", referring to figure 6-5.

7. No critical precision variable components. Easily automated by using transistor as the control element, therefore no mechanical or electromechanical components. Measurement value is in the form of a varying current (the current in the balancing loop) which can easily be transmitted over long distances.

8. By inserting the small voltage in the "galvanometer leg" of the bridge, replacing the variable resistance element with a fixed value precision resistor. Compare figures 6-5 & 6-6.

9. The fact that when the bridge is balanced (actually measuring the value of voltage) the current through the voltage source is zero. Therefore this is effectively an "infinite impedance" measuring system.

10. Refer to the figure following problem #16. The open circuits at points F and H, and C,N and D would be switched to change between resistance and voltage measurements.

11. Refer to the section on "Three-Wire Resistance Measurements".

12. Refer to the section on "Four-Wire Resistance Measurements".

13. Refer to the section on "Unbalanced Bridge Measurements".

14. The null-balance technique produces a condition in which several (two in a Wheatstone Bridge) signals are compared to produce a result that is essentially zero at the time the measurement is actually made. A second variable in the circuit is used to indicate the value of the measured variable.

15. Points H & J (points F,B & C will be at the same potential as point H for the resistance measurements).

16. The bridge circuit is relatively insensitive to nominal variations in the supply voltage. This is on of the big advantages of the Wheatstone Bridge measurement technique.

17. Connect points C and D (point N left open) and insert the millivolt potential between points F and H. The positive lead connected to point H and the negative lead to point F.

18. Connect points F and H together and insert the variable resistance element between points C and N. Point D is left open.

19. To give the bridge the capability of measuring several ranges of resistances and millivolt potentials.

20. The dc current (from the 100v power supply) through either the left or the right legs of the bridge will be:

100v / 100kohm = 1.0ma.

Therefore the voltage at point F will be: (50ohm) (1.0ma) = 50mv.

The voltage at point K will be: (30 + 20 ohms) (1 ma) = 50mv.

If the current in the current balance loop could be made zero then the minimum voltage that could be measured (between points F & H) would be zero. However the minimum value of current in the current balance loop will be: (1.0v) / (10 Kohm) = 0.1ma. This will cause an additional voltage drop across the 20 ohm resistor of: (0.1ma) (20 ohm) = 2.0mv. Therefore the minimum voltage at point K will be: 50mv + 2mv = 52mv. The range switch is connecting points J & K.

Therefore the minimum value of voltage between points F & H will be 52mv at point K minus 50mv at point F; or 2mv.
The maximum value will be with the maximum current flowing in the current balance loop (which will result in 1.0v drop across the 20 ohm resistor, with the 10K pot set to zero ohms). Assume the

resistance of the ammeter is adequate to drop negligible voltage across it. Therefore the maximum voltage at point K will be:

1.0v + 20mv + 30mv = 1.050v.

Each 50 ohm incriment of the range switch will add in:

(50 ohm) (1ma) = 50mv.

Since a total of 20 of them can be added in series with point K, then the maximum voltage at point J will be:

(1.050v at point K) + (20) (50mv) = 1.050v + 1.000v = 2.050v.

Therefore the maximum value of voltage that can be measured is:

(2.050v at point J) - (50mv at point F) = 2.000v.

21. Referring to the logic and calculations of problem #20, the minimum value of voltage at point J will be 52mv. The maximum value of voltage at point J (maximum current flowing in the current balance loop plus all range resistors added in) will be 2.050v.

Therefore the minimum value of resistance that can be measured must cause a 52mv potential at point C (and therefore at points F and H). This is: (52mv) / (1ma) = 52 ohms.

The maximum value will cause a 2.050v potential at the same points. This will be: (2.050v) / (1ma) = 2,050. ohms.

This maximum value assumes that the 2.050v at point C will have negligible effect on the 1ma of current which flows in that leg. In fact the 1ma will drop off to approximately:

(98v) /(100K ohms) = 0.98ma.

Therefore the actual value will be approximately:

(2.050v) / (0.98ma) = 2,092. ohms.

This will introduce nonlinearity into the measurements at higher values of resistance.

22. Since the circuit draws essentially no power from the source it is an extremely HIGH impedance measurement system.

23. The technique illustrated in figure 7-18 is very common.

24. Current through the leg containing the null meter.

25. This method draws essentially no power from the unknown signal voltage source.

26. The need to compensate for lead resistance.

27. Use the pulses to trigger a calibrated pulse generator circuit and integrate those calibrated pulses (with time) across a large capacitor.

28. It takes little, if any, current or power from the measured signal source; is quite susceptible to electronic noise; is specifically designed to reduce "meter loading" problems.

29. Because the resistance changes are only a few percent of the total resistance; due to the extreme self-heating effect of the strain gage itself; because of the very small (and noise prone) electrical output.

30. Approximately 1,000.mv.

31. Approximately 2.0mv.

32. E.

33. Refer to the section on "Origin of Grounding Problems".

34. Refer to the section on "How to Recognize a Grounding Problem".

35. - Use battery operated test equipment
- Use DC-to-DC coupling techniques
- Use good grounding practice in equipment design
- Use good grounding practice in equipment construction
- Use good quality power transformers
- Use good quality power supplies

36. I_3 = 14.75ma. $I_1 = I_2$ = 1.0ma from the text.
V_{R5} = (10 ohm) (14.75ma + 1ma) = 157.5mv
V_{R4} = (20 ohm) (1ma) = 20mv
$V_b = V_{R4} + V_{R5}$ = 157.5mv + 20mv = 177.5mv = V_a

$R_{unk} = V_a / I_1$ = 177.5mv / 1ma = 177.5 ohms

37. If I_3 = 4ma then V_{R5} = (4ma + 1ma) (10 ohm) = 50mv
If I_3 = 20ma then V_{R5} = (20ma + 1ma) (10 ohm) = 210mv
V_{R4} = (20 ohm) (1ma) = 20mv in either case
Max V_b = 210mv + 20mv = 230mv = Max V_a
Min V_b = 50mv + 20mv = 70mv = Min V_a

Range of R_{unk} = (230mv) / (1ma) = 230 ohm Max
= (70mv) / (1ma) = 70 ohm Min

38. V_a when R_{unk} = 1000 ohm = (1000 ohm) (1ma) = 1000mv = V_b
$V_a = V_{R4} + V_{R5}$ = (1ma) (20 ohm) + (20ma + 1ma) (R_5)

R_5 = (1000mv - 20mv) / 21ma = 46.67 ohm => 47 ohms (approx)

39. With I_3 = 0ma ;
V_b = (20 ohm) (1ma) + (47 ohm) (1ma) = 67mv = V_a
Therefore R_{min} = (67mv) / (1ma) = 67 ohms

With I_3 = 4ma ;
V_b = (20 ohm) (1ma) + (47 ohm) (4ma + 1ma) = 255mv = V_a
Therefore R_{unk} = (255mv) / (1ma) = 255 ohms

40. Refer to the text associated with figure 7-12.

41. A. 42. E. 43. A. 44. E.

Chapter 8

1. The term is accepted as describing just about any electronic amplifier having the following characteristics:
   - extremely high input impedance
   - extremely high gain
   - very low input impedance
   - dc coupling

2. Electronic simulation and solution of mathematical equations.

3. Just about any conceivable electronic amplifier application.

4. Due to the extremely high industrial and commercial use of these devices; also due to their extreme versatility.

5. The op amp as purchased, having the characteristics as listed in #1 above, before the user adds any external feedback elements.

6. The voltage at the summing junction (inverting input) of an op amp, when being used in it's inverting configuration, will be so small that it will be virtually at ground potential (zero volts).

7. Normally in a single integrated circuit, designed for general application rather than specific applications, with the characteristics outlined in question #1.

8. Means that the amplifier draws very little power from the preceding circuit, it does not "load down" the previous circuit.

9. Means that loading down the output will have relatively little (if any) effect on the value of the output voltage from the amplifier.

10. Power supply ground, plus and minus connections (if all used), frequency and temperature compensation terminals, zero (or offset) adjust terminals, the noninverting input if it is not being used.

11. Feedback is normally applied to the inverting input but not to the noninverting input. A voltage applied to the inverting input will be amplified at the output with polarity inversion, whereas a voltage applied to the noninverting input will be amplified at the output with the same polarity.

12. There are two different input terminals, one for the open loop amplifier and a different input for the closed loop amplifier. The input impedances will differ, the overall circuit gains will differ, the gain stability will differ, and the bandwidths will differ.

13. A. To stabilize the gain and to set the gain lower than the open-loop gain.
   B. Provides the closed-loop input impedance.
   C. $- R_{fb} / R_{in}$.
   D. A ratio of 100:1.
   E. Gain stabilization, gain reduction, bandwidth improvement, reduction in input impedance.
   F. It is absolutely NOT ground. If it was actually zero volts the output would go to zero regardless of the input voltage or the circuit gain.

14. A. It would provide positive feedback, and the circuit would saturate.
   B. It will always equal the output voltage divided by the gain of the amplifier.
   C. Closed-loop gain = $+ (R_{fb} + R_{in}) / R_{in}$.
   D. The input impedance of the op amp itself.

15. It remains at its same value, it is built into the op amp and cannot be changed.

16. Rephrase the section on "Op Amps As Differential Amplifiers".

17. Rephrase the appropriate portions of "Other Op Amp Parameters".

18. Rephrase the appropriate portions of "Other Op Amp Parameters".

19.

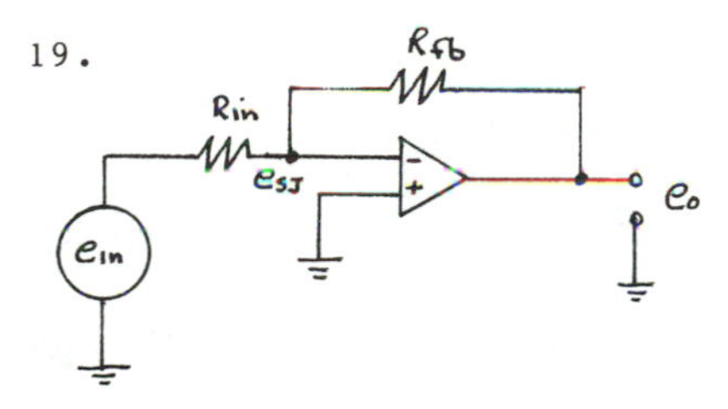

Closed-loop gain = $-R_{fb}/R_{in}$
also = $-e_o/e_{in}$

Open-loop gain* = $-e_o/e_{sj}$

20.

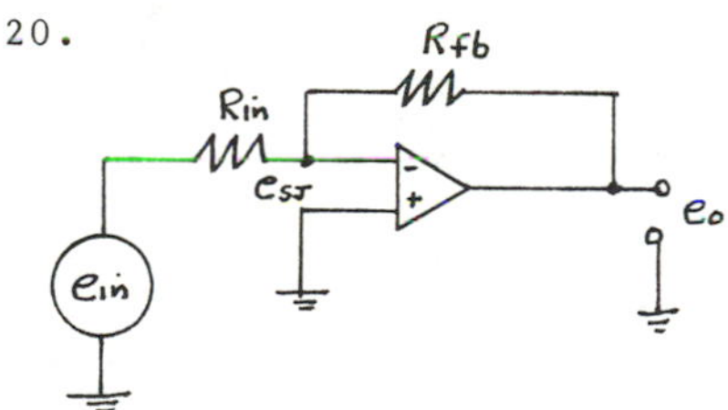

Closed-loop gain = $(R_{in}+R_{fb})/R_{in} = e_o/e_{in}$

Open-loop gain* = $e_o/(e_{in}-e_{sj})$

* Note that open-loop gain cannot be expressed as a ratio of resistors.

21. Plus, minus and ground power supply terminals (if in fact all three are used), frequency compensation terminals, temperature compensation terminals, zero-adjust terminals, and the

noninverting input terminal if it is not being used (in that case it would be grounded, probably through a resistor).

22.

$$e_o = -\left[-2v\left(\frac{100k\Omega}{20k\Omega}\right) + 1.25v\left(\frac{100k\Omega}{100k\Omega}\right) + 33v\left(\frac{100k\Omega}{300k\Omega}\right)\right]$$

$$= +10v - 1.25v - 11v = -2.25v$$

23.

$$\frac{e_o}{e_{SJ}} = -\text{Open loop gain} \quad \therefore e_{SJ} = \frac{-e_o}{\text{Open loop gain}} = -\left[\frac{-2.25v}{10^5}\right] = +22.5\mu v$$

24.

$$Z_{in_1} = R_1 = 20k\Omega$$

$$Z_{in_2} = R_2 = 100k\Omega$$

$$Z_{in_3} = R_3 = 300k\Omega$$

25.

$$I_1 = \frac{2v}{20k\Omega} = 0.1ma$$

$$I_2 = \frac{1.25v}{100k\Omega} = 0.0125ma$$

$$I_3 = \frac{33v}{300k\Omega} = 0.11ma$$

26.

$$e_o = e_{in}\left[\frac{R_{fb} + R_{in}}{R_{in}}\right] = +1.5v\left[\frac{100k\Omega + 50k\Omega}{50k\Omega}\right] = (1.5v)(3) = +4.5v$$

27. Since e_o=(open-loopgain)(e_{in}-e_{sj}) then e_{sj} must be approximately equal to e_{in} in both magnitude and polarity, or

$$e_{sj} \approx e_{in} = +1.5v.$$

28. With $e_{in} = e_{sj}$ (both are shorted together electrically), any voltage (except zero offset voltages) appearing at the output as the input varies will be a common-mode error.

29. Z_{in} of the op amp.

30. If the 50Kohm resistor were removed from the circuit then R_{in} would essentially be an infinite value. If the 100Kohm resistor were shorted out then R_{fb} would essentially be zero.

Therefore, from the gain equation, gain would be; $-\frac{\infty + 0}{\infty} \approx 1$

where the infinity symbols represent extremely large values.

31. $e_o = (e_{in} - e_{sj})$.

32. Nearly the same as e_{in}.

33. The presence of an output voltage due to the fact that both e_{in} and e_{sj} are at some voltage above ground.

34. $(R_{fb} + R_{in}) / R_{in}$

35. The same as the open-loop impedance.

36. Exactly the same as e_{in}.

37. The amplifier would saturate and stay there.

38. E.

39. Gain stability improved, sensitivity to component variations improved, frequency response reduced, gain reduced.

40. A. 41. 250ohms. 42. 100mw. 43. 250ohms.

44. 2.5 μw.

45. Saturates easily, nonlinear output.

46. A. 47. D.

48.

$$e_o = \frac{R_{fb} \to 0 + R_{in} \to \infty}{R_{in} \to \infty}$$

Gain approaches $\frac{0+\infty}{\infty} \approx 1$

Unity gain amplifier is commonly used as a noninverting buffer or isolation amplifier.

Chapter 9

1. Possible examples would include: the learning process, living, financial affairs, world affairs, etc.

2. Refer to figures 9-1,2,3,5,6,& 7.

3. So that the value of the output voltage is essentially not affected by loading.

4. To reduce the effects of extraneous noise, other than that the op amp is really not as necessary for differentiation as it is for integration.

5. Noise.

6. A = input waveform.
   B = integrator output waveform.
   C = differentiator output waveform.
   D = gain of less than one.
   E = gain of greater than one.

7. See definition of Analog Computer in the glossary.

8. Simulation and solution of mathematical equations (and non-mathematical processes which can be expressed mathematically).

9. See definition of Analog Computer Program in the glossary.

10. Refer to appropriate portions of "Analog Computers" section.

11.

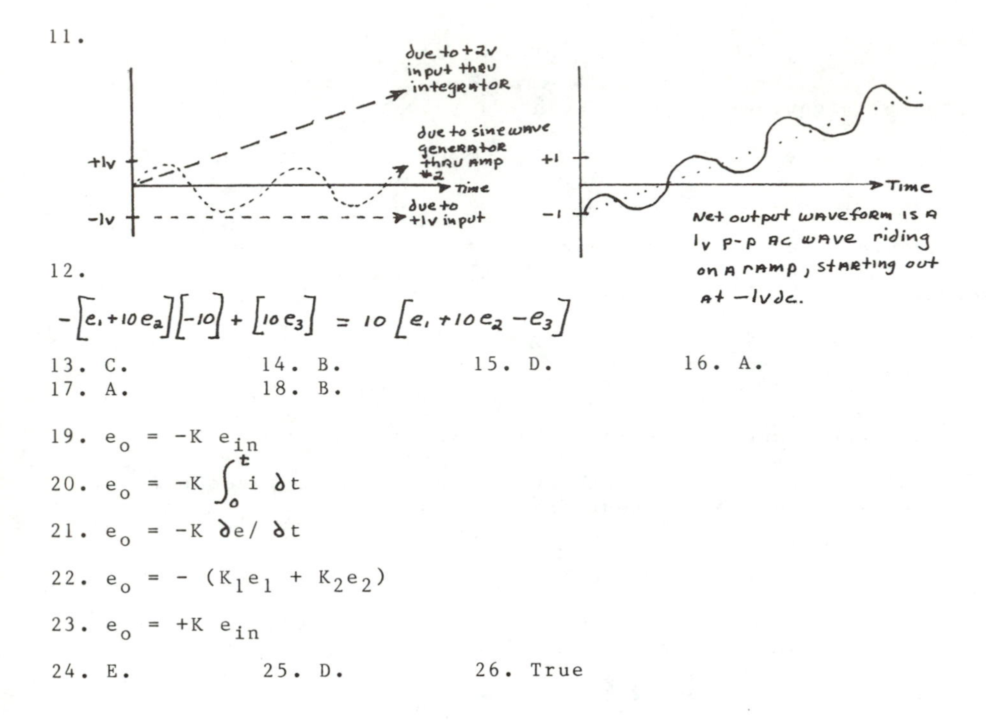

12.

$$-\left[e_1 + 10e_2\right]\left[-10\right] + \left[10e_3\right] = 10\left[e_1 + 10e_2 - e_3\right]$$

13. C. 14. B. 15. D. 16. A.
17. A. 18. B.

19. $e_o = -K\ e_{in}$

20. $e_o = -K \int_0^t i\ \partial t$

21. $e_o = -K\ \partial e / \partial t$

22. $e_o = -\ (K_1 e_1 + K_2 e_2)$

23. $e_o = +K\ e_{in}$

24. E. 25. D. 26. True

Chapter 10

1. Refer to the introduction.

2. Again refer to the introduction.

3. Refer to the first paragraph of the section on "Alarm Units".

4. Input variable = Temperature (from the thermocouple).
   Signal conditioning = Bridge circuit and amplifier (and probably a filter).
   Comparator = Op-amp-type circuit.
   Logic = Probably a relay-type circuit (possibly solid state).
   Switching element = Output of the relay circuit.
   Alarm = Siren (including any necessary drive circuitry).

5. Input variable = Temperature (from the RTD).
   Signal conditioning = Bridge circuit and amplifier (and probably a filter).
   Comparator = Op-amp-type circuit.
   Logic = Probably a relay-type circuit (possibly solid state).
   Switching element = Output of the relay circuit.
   Alarm = Claxon (including any necessary drive circuitry).

6. Refer to various figures in this chapter for examples.

7. Refer to figure 10-4 as an example.

8. Thumb through for yourself, they're there.

9. Refer to the section on "Recording And Indicating Equipment".

10. Refer to the section on "Transmitters".

11. See listing on page 241 in the text.

12. It linearizes the output from a sensor. Various pressure and flow sensors have square law outputs.

13. Refer to the section on "Transmitters" and also refer back to chapter 6, the section on "Frequency-Type Measurements".

14. Refer to the section on "Telemetry".

15. Garage door openers, citizens band radios, walkie-talkies, pagers, etc.

16. D. 17. A. 18. True. 19. A.
20. True.

Chapter 11

1. See definition of Controller in the glossary.

2. See the section on "Uses Of Controllers" in this chapter.

3. In a word "feedback".

4. A controller which has the ability of exercising ONLY proportional control, as further defined in the glossary.

5. Refer to the appropriate illustrations in figures in figures 11-2 and 11-3.

6. Summarize figure 11-4, and the example systems discussed for figures 11-2 and 11-3.

7. Controller gain-to-system stability relationship and acceptable controller gain-to-process output offset relationship.

8. AS the controller gain increases beyond a certain optimum value the system stability becomes adversely affected.

9. To reduce the offset inherent in proportional-only control to zero.

10. Described in the curves in figure 11-7. Generally improved long term control but reduced response to transients.

11. To improve controller response to transient conditions.

12. Described in the curves in figure 11-10. Generally improved transient response and some improvement in system stability.

13. Described in control of process step #2 of figure 11-13 (by controller #2).

14. Cascade control is described in control of process step #3 of figure 11-13 (by controllers #4 and #3). Also the setpoint can be supplied by a digital computer (called setpoint control).

15. The system "Capacitance".

16. Adjustment of the relative gains for the modes of control, with the controller actually on-line in the process.

17. The digital computer is actually in the feedback control loop of each process variable, and it is acting as the controller.

18. A.

19. Proportional, Integral and Derivative.

20. B. 21. B. 22. B.

23. Proportional plus derivative.

24. Derivative (pre-act or rate).

25. Integral (reset).

26. D. 27. E. 28. Derivative. 29. Integral.

30. Derivative. 31. Integral. 32. True.

Chapter 12

1. See definition of Controller in glossary.

2. Capability to exercise control over a single loop including setpoint and feedback inputs, command output, displays of current process variable's value and setpoint, manual and automatic control modes (at least), and gain control over each of the modes of control in the controller.

3. See appropriate portions of "Analog Electric Controllers" in this chapter.

4. See appropriate portions of "Analog Electric Controllers" in this chapter.

5. The only differences would be in the modes of control included (P, I & D) and in the relative gains set for each of the modes; otherwise the controllers do not differ at all.

6.

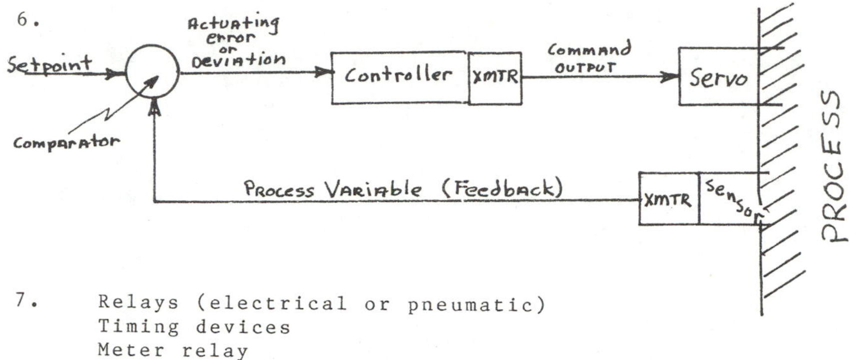

7. Relays (electrical or pneumatic)
   Timing devices
   Meter relay
   Cam controller
   Drum sequencer (or drum controller)
   Fluidic logic controller
   Programmable logic controller (PLC)
   Microprocessor controller
   Microcomputers
   Minicomputers
   Medium-size computers
   Large computers

8. See definition of programmable controller in the glossary.

9. Drum controllers must sequence through a fixed series of program steps, while a programmable logic controller has the ability to alter its own program steps based upon process conditions. New PLC's also include analog input/output capability which is impossible with drum controllers.

10. Primarily by cost, processing capability and physical size.

11. Primarily minicomputers, however now microcomputers are becoming more popular and common (because their cost has been decreasing thereby opening new markets for them). The medium size computers are used in larger processes.

12. Refer to the section on "Pneumatic Controllers" and figure 12-14 in this chapter for a description of flapper-nozzle operation.

13. Acquisition cost (pneumatic controllers are less expensive), process constraints which might dictate one over the other (such as explosive atmosphere), required speed of response (electronic controllers are faster), possible interfacing requirements with electronic computers, availability of maintenance and repair personnel already trained in one or the other, etc.

14. A digital computer which is made up using pneumatic digital logic elements exclusively.

15. Explain figure 12-18.

16. A machine tool whose operations are controlled by a programmable logic controller (electronic or pneumatic).

17. E.

18. A device which performs the same function as any electronic meter, monitors a process variable and actuates a relay when the measured variable exceeds or falls below preset values.

19. E.

20. Percent of full scale value.

21. Every piece of control system equipment except sensors and servos (and of course the process itself).

22. There is basically no difference in the physical controllers.

23. E.	24. D.	25. E.	26. 4.
27. N.	28. M.	29. 1.	30. 2.
31. 3 & 4.	32. O & P.	33. 3.	34. A.
35. E.	36. E.	37. D.	38. E.

39. E.

40. A relay ladder diagram.

Chapter 13

1. When digital computers were first applied to process control and the need for an interface between the analog process and the digital computer developed.

2. Number of bits of conversion resolution.

3. Resolution in A/D or D/A conversion systems refers to the degree of fineness of detail in the output, it is a measure of the smallest possible incrimental change in the output due to a corresponding change in the input (not noise). This is a measure of the quantizing error in the device. If the valid change in input is compared to the resulting change in the output, then you have a measure of the sensitivity. Resolution implies both quantizing error and sensitivity in A/D and D/A conversion systems.

4. A ratio of the full scale input (volts) as compared to the full scale output (in maximum number of possible combinations of bits). The dimensions are therefore volts-per-bit.

5. This is one of the basic techniques used in commercial D/A conversion systems. Rephrase the appropriate portions of "Digital-to-Analog Conversion Techniques", referring to figure 13-1 as necessary.

6. Rephrase appropriate portions of the section on "Linear Ramp Encoder Technique", referring to figure 13-6 as necessary.

7. Rephrase appropriate portions of the section on "Dual Slope Integrator Technique", referring to figures 13-7 and 13-8 as necessary.

8. Advantages of dual slope over linear ramp techniques:
- conversion accuracy is not dependent upon critical specifications of the integrator components.
- long term variations in the basic clock frequency will not affect conversion accuracy.
- more imune to electrical noise.
- can economically be made to be more accurate.

Disadvantages:
- dual slope is more complex in circuitry, design and understanding.
- dual slope is more costly to manufacture.

9. Rephrase appropriate portions of the section on "Successive Approximations Technique", referring to figure 13-9 as necessary.

10. Advantages of linear ramp over successive approximations:
- less complex in design and understanding.
- less costly to manufacture.

Advantages of successive approximations over linear ramp:
- faster.
- more adaptable for use with digital computers.

Neither enjoys an advantage in-as-far-as conversion accuracy is concerned.

11. Basically it is a filtering function. It converts a varying voltage to a non-varying dc voltage so that the A/D converter will accurately make the conversion; so that the voltage to be converted will not change during the actual conversion process.

12. See definition of multiplexing in the glossary, and also refer to figure 13-16.

13. The ability to use a single A/D converter (or other strictly digital element) for a multitude of purposes by selecting which of several inputs are connected at any given period in time.

14. An amplifier which has a selection (digitally selected) of gains available. In an A/D conversion system it takes several different ranges of input voltages and amplifies them up to the level required by the A/D converter following it.

15. Refer to figure 13-17.

16. The flip-flop register is universally used for this application.

17. In the number of bits of resolution.

18. The resolution of this 5-bit system (which is obviously a full 5 bits of resolution, not 4 bits of resolution plus a sign bit), will be:

$$\frac{10v}{2^5 \text{ bits}} = \frac{10v}{32 \text{ bits}} = 0.3125 \text{ v/bit}$$

Therefore the output voltage will be 31 (which is the decimal equivalent of 11111 binary) times the scale factor 0.3125 v/bit or it equals 9.6875v.

Notice that it is impossible to get the full scale output value. The closest that the system can approach the full scale value is within one incriment of resolution.

19. The resolution of an 8-bit, 10-volt system is $10v/2^8$ bits, which equals 0.039 v/bit. The sensitivity is therefore 0.039 volts; in other words the minimum change in input voltage which will produce a detectable change in the output is 0.039 volts.

20. $10101_2 = 21_{10}$. Therefore the voltage is 21 times 0.01v/bit which equals 0.21 volts. Obviously the full scale output from this A/D converter is 0.32 volts.

21. #1 = comparator
#2 = D/A converter
#3 = Flip-flop register
#4 = Reference power supply

22. To make a pre-programmed guess and then refine it based upon internal feedback as to the validity of that guess.

23. Digital ramp encoder technique.

24. #1 = Linear ramp generator
#2 = Comparator
#3 = Counter
#4 = Oscillator

25. B.

26. #1 = Switch
#2 = Integrator
#3 = Comparator
#4 = Counter
#5 = Clock

27. A.

28. "In front" of the A/D converter.

29. Flip-flop registers.

30. Any inaccuracies generated by the integrator cancel out.

31. An electronic device which converts the value of a continuous signal into a discrete code.

32. Reference voltage source, comparator, digital parallel output register and source of digital timing pulses.

33. E.

34. A very stable frequency digital clock, a binary counter, an op-amp comparator and some digital logic.

35. B.

36. Approximately 0.01v

37. Approximately 0.1v

38. It is not possible to get exactly 10.v out of this D/A converter; the maximum will be (10.v - 1 LSB).

39. 5 bits. 40. E. 41. A.

Chapter 14

1. Rephrase appropriate portions of the section on "What A Digital Compute Is", also include what it is not.

2. Rephrase appropriate portions of the section on "How Digital Computers Exercise Control".

3. See definition of the word data in the glossary.

4. See definition of the word "word" in the glossary.

5. See definition of the word binary in the glossary.

6. Logical operations refer to operations performed on binary-coded data which can be described by the rules for Boolean algebra.

7. Supervisory controller or computer refers to the application of a computer whose function is to provide the setpoints to other individual controllers; it does no direct control itself, the other controllers are actually exercising the control function.

8. See definition of the word peripheral in the glossary.

9. See definition of the term central processing unit in the glossary.

10. See definition of the word register in the glossary, also refer to the definition of buffer register in the glossary.

11. A flip flop is a bistable digital electronic circuit which has the capability of temporarily storing one binary bit of information.

12. Mass memory refers to digital computer memory devices designed to provide auxiliary storage of large quantities of digital information in support of the computer's main (working) memory.

13. See definition of instruction fetch cycle in the glossary, also refer to figure 14-2 and the supporting text material.

14. See definition of the word program in the glossary.

15. Refer to appropriate portions of "How A Digital Computer Differs From An Analog Computer" and "How A Digital Controller Differs From Other Controllers".

16. - Multiplexing of single computer elements to all process control loops rather than having to duplicate a complete set of hardware for each loop.
- Large scale and permenant memory (or at least storage for an unlimited period of time).
- More complex control system computations possible.
- Additional control schemes, such as adaptive and optimal.
- Drifting of components with time, temperature, etc. has no inherent effect on the basic accuracy of the system, therefore no routine recalibration procedures are necessary.
- System can be programmed to monitor its own performance, reporting system failures instantaneously, even taking corrective action if so programmed.
- Input/output displays and controls with performance capabilities far in excess of those achievable in analog systems.
- Easily altered, modifiable control system strategy and control of parameters without changes to hardware.
- System more economically expandable.
- System available for non-process control functions, such as production scheduling and inventory control.

17. Refer to appropriate portions of "How A Digital Controller Differs From Other Controllers".

18. Refer to appropriate portions of the introductory paragraphs for this chapter.

19. Refer to appropriate portions of "Types Of Digital Computers".

20. Refer to appropriate portions of "Types Of Digital Computers".

21. Rephrase appropriate portions of the section on "How Digital Computers Work".

22. Refer to figure 14-1 and describe the diagram.

23. See definition of program address register in the glossary.

24. See definition of memory address register in the glossary.

25. See definition of memory data register in the glossary.

26. See definition of instruction register in the glossary.

27. Refer to appropriate portions of "How Digital Computers Work" and also refer to figure 14-2.

28. Matrix printers; modems; network interfaces; cathode-ray-display systems; cassette, cartridge, reel-to-reel tape, floppy disc and hard disc memory systems; A/D and D/A conversion systems (including all of the equipment necessary to operate them); line printers; etc.

Chapter 15

1. See definition of computer architecture in the glossary.

2. Summarize appropriate portions of the section on "Information Required By The Computer", and figure 15-1.

3. Summarize appropriate portions of the section on "Information Required By The Process", and figure 15-2.

4. See definition of multiplexing in the glossary, figure 15-3, and appropriate portions of the section on "Computer Input Bus".

5. See definition of the word interrupt in the glossary, and appropriate portions of the section on "Computer Input Bus".

6. Device addressing serves the purpose of selection of one out of several specific individual devices which are multiplexed together, or selection of specific devices in a demultiplexing network.

7. A parallel adder, two´s compliment arithmetic capability, and shifting capability.

8. Interface is the term used to describe the total quantity of equipment required to provide electronic plug-in capability between the digital computer and the process (and peripherals).

9. Refer to figure 15-1 and the associated text material which refers to it.

10. Refer to figure 15-2 and the associated text material describing that figure.

11. To provide the digital computer with a time reference in order to perform its time-based functions (ie. convert a specific analog process variable´s value to digital and run the process control algorithms for that loop every 100ms).

12. Sequentially controlling each internal operation of the computer.

13. See definition of the term machine language programming in the glossary.

14. In the computer's main memory; typically RAM memory or a combination of RAM and ROM.

15. The computer cycle is made up of several sequential events including (but not necessarily limited to, or necessarily in this order) the instruction fetch cycle, the time required to actually decode and execute a basic one-cycle instruction, the time required to check for any interrupts that may have occurred, the time required to check for direct memory access requests, etc. This sequence of events is repeated (at least once) for each instruction.

16. See definitions of the terms compiler and machine language programming in the glossary.

17. Refer to the answer to question #16 above plus definitions of the terms assembler and assembly language programming in the glossary.

18. Rephrase the "Summary" for this chapter.

19. Because the internal operations are controlled by the instructions, which are internally represented in their binary form; regardless of what language the program was originally written in. Also, in order to troubleshoot, repair and/or test computers or their peripherals; their software drivers are normally programmed directly in machine language.

20. Figure 15-6 is a functional block diagram of a digital computer, also it illustrates the functional connections for both the input and the output interfaces, and the addressing scheme.

21. Because, through the computer, the same set of computational logic hardware can be multiplexed to every control system loop, whereas with hardwired logic the logic circuits must be duplicated for each loop.

22. C. 23. E. 24. D. 25. A.

26. C. 27. E. 28. D.

29. To cause the computer to immediately and unconditionally deal with unusual conditions (either emergency or seldom occurring but important) or to deal with routine situations on a "on demand" basis, and also they can be used to provide sycnhronization between asynchronous devices.

30. True 31. C.

32. The digital computer receives all sensor inputs directly, performs all control system calculations, and provides the control signals directly to the system actuators and indicators.

33. The computer requires standardized digital information which is normally not electronically compatible with process transducers.

34. The same logic circuits in the computer can be time-shared by many separate control system functions, and need not be physically duplicated, nor rewired when changes in the control system algorithms are necessary.

35. Addressing. 36. C. 37. D. 38. C.

39. E. 40. A.

41. To obtain the next computer instruction to be executed and insert it into the instruction register.

42. B. 43. D. 44. C. 45. A.

46. Word size. 47. E. 48. B.

49. A binary coded computer word which will be decoded by the instruction register logic and cause the computer to perform some useful and repeatable operation.

50. C.

51. Only a single data operation.

52. 8 or 16.

53. 8 or 16, but some more modern mini-microcomputers are based upon 32 bit word sizes.

54. Two´s compliment arithmetic.

55. The ability control the flow of data down alternative buses.

56. Addition, and sometimes multiply and divide.

57. The accumulator is used for mathematical operations also.

58. Computer programming.

59. A computer program.

60. Sequentially.

61. Branch instructions and conditional instructions.

62. The gating between data registers, accumulators, memory, etc. (all data flow internal to the CPU).

63. See answer to #62 above, plus to "rewire" the internal computer logic for multiple purposes, decode the various operation codes, and to control all data transfer internally and external to the computer.

64. The program counter.

65. The preceding operation defines the following word as data, instruction, etc.

66. To keep track of where the next instruction to be executed will come from.

67. Testing the carry bit (or flag).

68. A.

Chapter 16

1. See the definition of actuator, and also the definition of transducer, in the glossary.

2. See the definition of the word servomechanism in the glossary.

3. The command output of the controller has been calculated and conditioned to drive the actuator, which actually causes the controlled variable (the feedback variable) to change.

4. Both are transducers and both are physical pieces of hardware. The difference is in the system function; sensors transduce system variables into a form for input to a controller, whereas actuators receive information from the controller and transduce that information into mechanical motion. The sensor is a control system input device whereas the actuator is a control system output device.

5. Refer to appropriate portions of the section on "Direct-Current Motors", and also figure 16-5.

6. To limit the initial inrush currents when first starting the dc motor, until it is rotating at a high enough speed where the counter-emf will be adequate to safely control armature current.

7. Controlling armature current, controlling field strength, and controlling load (refer to figure 16-8).

8. By reversing the supply polarity (refer to figure 16-9).

9. Refer to appropriate portions of the section on "Alternating-Current Motors", and figures 16-10,11 & 12.

10. To get the armature started, otherwise there will be no starting torque.

11. Refer to figures 16-14 and 16-15, and the supporting text information.

12. A series-wound dc motor. If the motor has a split ring commutator and runs on ac current then it is a universal motor.

13. By controlling the frequency of the applied excitation, which is normally not practical. Therefore normally the speed of the ac motor is not controlled, however the output power take-off shaft is coupled to the motor by some type of a clutch arrangement; thereby some degree of speed control over the output shaft will be achievable.

14. A synchronous motor has a permenant magnet rotor whereas an induction motor has an electromagnetic rotor.

15. Reverse the power leads to either the starting windings or to the main running field windings to change directions.

16. In a dc motor the field is of constant magnetic polarity and the magnetic polarity of the rotor alternates. In an ac motor the magnetic polarity of the field alternates while the magnetic polarity of the rotor remains constant.

17. The mechanical construction of the motor itself and the physical arrangement of the windings on the motor; not the number of steps it has taken. Therefore the motors are frequently used without feedback in controlled systems.

18. Basically it is similar to the ac and dc motors; attraction of unlike magnetic poles and repulsion of like magnetic poles.

19.

PM	VR
Separate pairs of stator poles	Single pole piece with individual poles cut in it
Permenant magnet rotor	Rotor of magnetic material, not permenant magnet
Windings energized in parallel groups	Windings energized in a series arrangement
Detent characteristic with all windings deenergized	No detent characteristics
	Can be made economically with smaller step angles

Both: Windings energized sequentially, bidirectional rotation, stepping action, direct digital actuator.

20. Refer to figure 16-22, the diagram is fairly self-explanatory; however note the system power gain. It takes very little power to move the spool valve, and it will remain in any fixed position without requiring the application of any power to hold it there, whereas the power available from the piston is limited only by its size and the pressure available from the pump supplying the system. Therefore the power gain is extremely high and this accounts for the common use of this type of system (ie. automotive automatic transmissions, backhoes, dump trucks, plows, etc.).

21. Refer to figures 12-14, 12-15 & 16-25. Figure 16-25 is concerned more with the principle of operation and construction of the actuator itself, whereas the figures in chapter 12 are not

concerned as much with the actuator as they are with its inclusion in a system. Figures 12-14 and 12-15 use the diaphragm motor as the actuating element in a control system. In all three cases the diaphragm motor operates on the same principles and could even be the same physical piece of equipment.

22. See page 35 for the answer to this question.

23. D. 24. E. 25. B. 26. C.

27. Approximately the same pressure as the supply.

28. Regulation of output pressure.

29. D. 30. D. 31. E.

32. The presence or absence of feedback around it.

33. C. 34. E. 35. True.

36. Universal (dc) motors.

37. D.

38. Starting torque.

39. It improves the speed regulation.

40. Series wound, shunt wound & compound wound.

41. Counter emf.

42. Connect the motor to its power supply through a switch which will reverse the direction of current flow through it or provide the motor with two sets of windings, one for each direction of of rotation.

43. To provide initial starting torque.

44. Permenant magnetic rotors operate at synchronous speeds and normally do not develop much power; electromagnetic rotors operate as induction motors and can easily be made to develop much more power.

45. The armature´s tendancy to act as a generator.

46. E. 47. B. 48. A.

Chapter 17

1. Refer to the introductory paragraphs of this chapter.

2. Refer to the introductory paragraphs of this chapter.

Problem #22, Chapter 16

Actuator	Variable Controlled	Principle of Operation	Advantages	Disadvantages
Digital Solenoid	Digital Position	electrical-to-magnetic-to-mechanical	inexpensive, effective	digital only, mechanical probs, speed
Analog Solenoid	Analog Position	as for digital solenoid	inexpensive, effective	low power out, low frequency, mechanical
Relays	Electrical Power	as for digital solenoid	I/O isolation, high power amp, inexpensive	as for digital solenoid
Rotary/Linear/Rotary	Form of Mech. Pos.	gears or worm screw	inexpensive, effective, can be high power	mechanical problems
DC Motors	Mech Rotary Motion	electromagnetic interaction	high power, speed ctrl, direction control	DC power, brush/commutator assy, starters, mechanical
AC Induction Motors	Mech Rotary Motion	as for DC motor	AC power, starters not rq'd, high power, cheap	speed ctrl difficult, mech probs, starting winding & switch, commutator & brush
AC Synchronous Motors	Mech Rotary Motion	as for AC motor	extremely constant speed	low power, speed control difficult, gear heads
Universal Motors	Mech Rotary Motion	as for AC motor	AC or DC operation, speed control, inexpensive	low power, brush/commutator assy, mechanical
PM Stepper Motors	Digital Shaft Pos.	as for ac motor except sequential operation	detent torque, no feedback rq'd, precise position, direct digital	expensive, complicated control logic, limited position control & power
VR Stepper Motors	Digital Shaft Pos.	as for PM except VR rotor	smaller step angles than PM, otherwise the same	no detent torque, otherwise same as for PM
Hydraulic Cylinders	Analog Shaft Pos.	hydraulic/mechanical	high power out, reliable, rugged, small power out/size relationship	mechanical probs, leaks & seals, expensive, control valves & pump required
Diaphragm Motors	Analog Shaft Pos.	pneumatic/mechanical	as for hydraulic clyinder plus availability of pneumatic controllers	as for hydraulic cylinders

3. Refer to the section on "Preliminary Steps".

4. The first step. Refer to the section on "First Step: Understanding The System".

5. - Attend schools on the system
 - Read manuals
 - Operate the system in all of its modes
 - Spend time with designers of the system
 - Spend time with other operators of the system
 - Refer to other sources of literature on similar types of equipment and systems
 - etc.

6. The computer itself, because normally there is at least a minimal level of diagnostics built in to help and because normally the system can be operated in all of its modes of operation from the computer so that the particular failures can be isolated.

7. There really isn´t a difference in the basic approach, however the magnitude of the task may be significantly greater.

8. There are too many possibilities to include here.

9. C-B-E-A. 10. B. 11. B. 12. E.

13. B. 14. B.

Chapter 18

1. Refer to the definition of robot in the glossary, also refer to the section on "What Are Robots?"; then add the concepts which are more specifically associated with industrial robots.

2. They are basically the same, operate on the same basic principles and in many cases look similar to industrial robots. The differences are in the application. The application dictates different styles of construction, ruggedness in construction materials used, quality of joints and end effectors, power available from installed drive motors, and complexity of the control systems. In a word, cost.

3. Work more efficiently, for longer continuous periods of time, more economically and with more consistency. I´m sure you cam come up with additions to this list.

4. Tasks which require a high degree of consistency in manufacture, to closer tolerances and at higher production rates. Again I´m sure you can expand upon this listing.

5. (1) some production line assembly operations
orienting parts for further assembly
loading and unloading machines, conveyors, pallets, etc.
packing crates, boxes, etc.

(2) stuffing printed circuit boards

(3) operations in space and on other planets
operations on the ocean floor

(4) many welding operations
many spray painting and material coating operations
high speed assembly and inspection operations

(5) production assembly line personnel in automotive plants

(6) maintaining some machining sequences and tolerances

6. I´ll leave this one to you.

7. Rephrase the explanation on p.449.

8. Basically the type of robot used (most can only be programmed one way) and the application (some operations are too complex to program any other way).

9. They use all of the same concepts, hardware and controls discussed in previous chapters. They are basically a specific packaging of that technology and those concepts, which has found widespread application throughout industry.

10. In many ways robots are comparable to analog controllers and digital computers in their flexibility. They are a specific packaging that can be reprogrammed, many times, without physical modifications, to accomplish a wide range of tasks.

11. (a) workpiece would have 6 degrees of freedom
(X, Y, Z, Yaw, Pitch, Roll)

(b) the three motions would only extend the work envelope in those directions, not add any more degrees of freedom

(c) as for answer (b), still only 6 degrees of freedom

12. Work envelope is the physical, three dimensional space in which the robot can effectively work.

Appendix A

1. Refer to table A-1.

2. Ohm´s law, Kirchoff´s voltage and current laws.

3. Should be a series circuit (system) illustrating that the sum of the pressure drops around a closed system must total up to the applied pressure.

4. Pressure drop across a restriction in a pipe is proportional to the flow through the pipe times the effective resistance of the restriction.

5. Make sure that the heat flow is accounted for at each system junction, and that the <u>heat flow</u> is illustrated, <u>not</u> the temperature!

Appendix B

	32,768 16,384 8,192 4,096	2,048 1,024 512 256	128 64 32 16	8 4 2 1	DEC	OCTAL	HEX	BCD
1.	0,001,	011,0	11,10	0,111	5863	013347	16E7	INV.
2.	1,100,	110,0	00,11	0,011	52275	146063	CC33	INV.
3.	0,101,	010,1	01,01	0,101	21845	052525	5555	5555
4.	0,000,	111,1	00,00	1,111	3855	007417	0F0F	INV.
5.	0,101,	101,1	10,10	1,101	23469	055655	5BAD	INV.
6.			01	1011	27	x	x	x
7.	0111	1011	1001	1010	31642	x	x	x
8.	0111	0100	0110	1000	29800	x	x	x
9.		0001	1100	1100	460	x	x	x

10.

1 0 0 1 1 0 1 1	=	155	(check)
0 1 0 1 0 0 0 1	=	81	
1 1 1 0 1 1 0 0	=	236	

11.

0 1 0 1 0 0 0 1	=	(B)
1 0 1 0 1 1 1 0	=	one´s compliment
+ 1	=	plus one
1 0 1 0 1 1 1 1	=	two´s compliment
1 0 0 1 1 0 1 1	=	(A)
1 1 1 1 1 1	=	carrys
0 1 0 0 1 0 1 0	=	(A) - (B) = 74 (check)

The end carry has no valid meaning in the answer.

12. A binary one in the sign bit (for two´s compliment arithmetic) means a negative value, therefore (A) would have been a negative number since the MSB (the sign bit) is a one.

13. By the same logic as in question #12, (B) would have been a positive number.

14. Shift left multiplies by the base of the number system (two in the binary system). Right shift divides by the base of the number system.

15. Eight bits has 2^8 possible combinations, each of which can represent some unique discrete event or piece if information (256 decimal unique states, including the all zero state).

16. By the same logic as in question #15, a 16-bit binary word can represent $2^{16} = 65{,}536$ unique pieces of information.

PRENTICE-HALL, INC., ENGLEWOOD CLIFFS, NEW JERSEY 07632

0-13-054487-